Bibliographic information published by the German National Library:

The German National Library lists this publication in the National Bibliography; detailed bibliographic data are available on the Internet at http://dnb.dnb.de .

Imprint:

Print and binding: Books on Demand GmbH, Norderstedt Germany
ISBN: 978-3-668-03063-3

This book at GRIN:

http://www.grin.com/en/e-book/302739/vegetation-assessment-of-amri-mined-out-sites-for-potential-nickel-hyperaccumulator

Joshua Reambonanza, Felix M. Salas

Vegetation Assessment of AMRI Mined-out Sites for Potential Nickel Hyperaccumulator Species

GRIN Publishing

VEGETATION ASSESSMENT OF AMRI MINED-OUT SITES FOR POTENTIAL NICKEL HYPERACCUMULATOR SPECIES

Table of Contents

Rationale

Plants as a tool for environmental remediation is a basic idea which is practiced in old times and cannot anymore be traced to any particular source (Raskin *et al.*, 1997). Some plants can tremendously accumulate high levels of certain metal ions (Cullaj *et al.*, 2004). The ability of these plants to accumulate very high concentrations of metals from soils provides the basis for remediation of the contaminated sites where toxic metals are removed from contaminated areas thru extraction. The traditional approach of remediating a polluted and contaminated site is extremely expensive and can range from $600,000 to $3,000,000 depending on the extent of pollution, whereas Phytoremediation costs 20 times less expensive (Paz-Alberto & Sigua, 2013). Plant species are considered hyperaccumulators if it accumulates more than 1000 mg/kg (dry weight) or 100-fold greater than in those common nonaccumulator plants (Shaw, 1989; Baker, 1995; Salt *et al.*, 1995; Lasat, 2002)

Claver, Surigao del Norte in the Caraga Region is one of the places to have the biggest iron mineral deposit throughout Asia and has attracted many mining companies. It has at least three (3) mining companies of which Nickel is one of the main heavy metals being mined-out. Adnama Mining Resources Incorporated (AMRI) is one of the companies established in the area which is engaged in Nickel mining.

There were several heavy metals hyperaccumulator plant species already identified (Fernando *et al.*, 2014) and reported worldwide by scientists in several countries. The application of this remediation strategy (using these identified plant species) however, is limited by the climactic and geological conditions of the site to be remediated such as the temperature, soil type, and altitude; it is further limited by the availability of the metals in the soil for uptake (Schmoger *et al.*, 2000). Hence, this study has been conducted to provide insight for the rehabilitation of the mined-out areas of AMRI.

Objectives

The primary objective of the study was to identify potential nickel hyperaccumulator native plant species that can be used as a tool for soil remediation in mined-out areas of Adnama Mining Resources Incorporated (AMRI) located at Brgy. Urbiztondo, Claver, Surigao del Norte, Philippines. Specifically it aimed to (1) assess the biophysical characteristics of the mined-out areas of AMRI with special reference on Nickel, (2) evaluate the nickel content of tissue samples of selected native plant species as potential Ni hyperaccumulator and (3) recommend native plant species suitable for restoration of mined-out areas operated by AMRI.

Methods

Samples of soil and plant species were collected from the mined-out areas of AMRI at Brgy. Urbiztondo, Claver, Surigao del Norte, Philippines. After field surveys and visual observations, four (4) sampling sites were identified. Three of these sampling sites were in the mined out areas still having natural vegetation while one sampling site was in the area that was not yet mined during the conduct of the study which served as a reference.

The sampling sites were established in the mined-out areas where natural vegetation is still present. Sampling area was divided into four sampling stations each with 10 x 10 meters. Topographical locations of the established sampling sites are modeled in Figure 1a and Figure 1b. First sampling site was established at the summit of the mined-out area, second was established at the backslope, and the third was established at the footslope. The fourth sampling site was established at backslope in an area which was not mined-out a few kilometers from the three (3) sampling sites. Established sampling sites 1, 2, and 3 are located at Block 2 Mined-out Area of AMRI while sampling site 4 was established at Block 3 of AMRI.

Four composite soil samples in four different strata (0-20, 20-40, 40-60, 60-100 cm depth) at each established site were collected and placed separately in properly labeled plastic bags. Soil samples were subjected to standard methods and procedures of soil sample preparation of the AMRI Assay Laboratory.

After the preparation of the collected soil samples, it was then subjected to XRF method of analysis for total Ni analysis. Standard methods and procedures of the AMRI Assay Laboratory were followed for total nickel analysis.

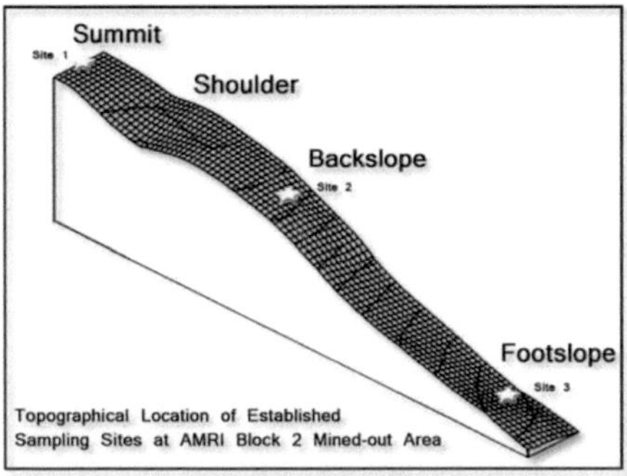

Figure 1a. Topographical Location of Established Sampling Sites in the AMRI's Mined-out area

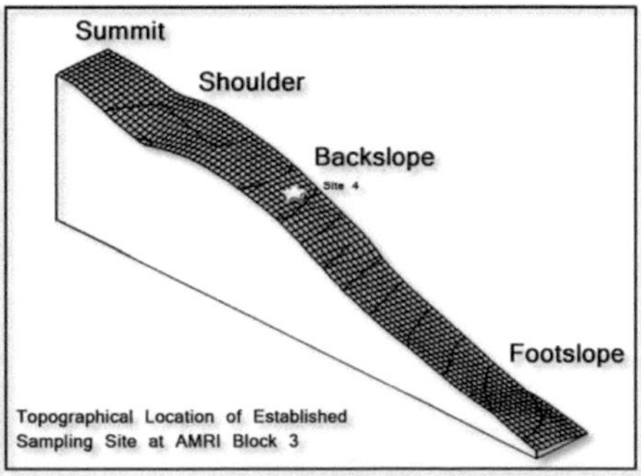

Figure 1b. Topographical Location of Established Sampling Site in the Block 3 (Not – Mined)

Healthy mature leaves of indigenous plants that were covered by the 10x10m established sampling site at each site were randomly collected. Leaf samples were taken from 5 different plants of the same species per sampling point to make a composite sample. In cases where the population of one species of plant was less than 5 in the sampling site, a composite sample was collected from the maximum number of plants present. Leaf samples were washed with distilled water in the field, air dried, kept inside labeled paper bags and was transported to the Department of Agronomy and Soil Science, Visayas State University. Samples were oven-dried at 70°C for at least 2 days before grounded and placed in labeled small paper bags.

Grounded leaf samples were screened (semi-quantitative test) for nickel accumulation. A 0.5 g of each grounded plant sample was wet digested with 6 M hydrochloric acid overnight. The digest were diluted to 100mL volume solution. Aliquots were taken and added with 1mL of 1% dimethylglyoxime dissolved in 95% ethanol. Formation of pink or magenta color indicated exceptionally high (above 1,000 μg g^{-1}) concentration of Ni in the dry plant matter (Fernando *et al.*, 2014).

Standard nickel previously added with 1 % dimethylglyoxime, dissolved in 95% ethanol was set-up for UV-Vis analysis. The wavelength of this sample served as the basis in qualitatively determining the plant samples for possible nickel hyperaccumulation.

Top ten (10) samples were considered from the 81 plant samples that had undergone semi-quantitative screening for nickel accumulation. These top ten (10) samples were then identified through the detailed photographs sent and examined by the AMRI's Rainforestation Officer, Mr. Cyril Reginald Canonoy. Other plant species which were difficult to identify were identified in the family level by Forester Hernando L. Mondal of the College of Forestry and Environmental Sciences in Visayas State University

The top ten (10) samples were then analyzed quantitatively for total nickel concentration using the UV-Vis Absorption Spectroscopy. Kept, grounded samples of these top ten were ashed at 600°C for 2 hours followed by the addition of 1mL concentrated hydrochloric acid. The digests were diluted and aliquots were taken then 1mL of 1% of the nickel-specific colorimetric reagent, dimethylglyoxime, dissolved in 95% ethanol was added. After which, samples were taken for nickel analysis through UV-Vis Absorption Spectroscopy. Standard methods and procedures were adapted from Salas *et al.*, (1998).

Levels of μg g-1 of Ni in leaf samples were compared to the criteria for classifying metal hyperaccumulator plant species. To be considered hyperaccumulating, the plants must have accumulated 1000 mg/kg or 1000 μg g^{-1} (dry weight) or 100-fold greater than in those common nonaccumulator plants of Ni (Shaw, 1989; Baker, 1995; Salt *et al.*, 1995; Lasat, 2002; Baker and Brooks, 1989).

Soil and Plant sample's Ni analysis results were subjected to statistical analysis (ANOVA) using STATISTICA software version 8.0.

Findings

Total Ni contents in soils of the four established sampling sites ranges from 7,600 μg g^{-1} – 9, 900 μg g^{-1} with a mean of 8, 200 μg g^{-1} (Figure 2). Statistical analysis of these results shows that there is no significant difference of total Ni content among the sites which entails that within (soil strata) and among the sites, the total Ni values are approximately the same.

Based on the XRF results, an area can be classified or categorized whether it's low or high grade. The lowest grade is 0.24% or 2,400 μg g^{-1} while the highest grade is 2.82% or 28, 200 μg g^{-1}.

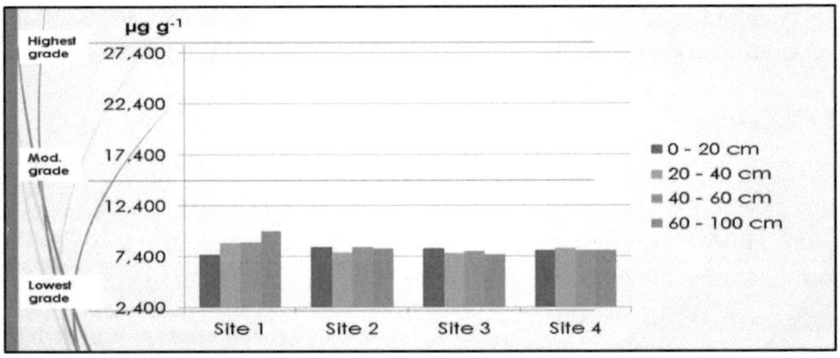

Figure 2 .Ni analysis (XRF) in soil samples results

There were 81 plant samples collected in the sites and was screened for nickel accumulation. Table 1 below shows the top ten (10) of the plant samples ranked according to their absorbance level after the semi-quantitative screening (Appendix 1) using UV-Vis Spectrophotometer.

Table 1. Top ten plant samples according to their absorbance levels after Ni accumulation screening

Plant Samples	Absorbance	Potential Rank
S1P13	0.6860	1st
S2P11	0.3094	2nd
S2P14	0.2871	3rd
S4P3	0.2754	4th
S1P3	0.1573	5th
S4P5	0.1560	6th
S4P16	0.1460	7th
S1P18	0.1354	8th
S3P7	0.1350	9th
S4P22	0.1260	10th

Kept, grounded samples of these top ten plant samples were then analyzed quantitatively for total nickel concentration using the UV-Vis Spectrophotometer. Table 2 shows the amount of total %Ni in the Leaf Samples of Plants selected after Ni quantification of the screened plant samples.

It was found that one of the plant species identified belonging to the Rubiaceae family and is locally known as *"Sowa-sowa"* had the highest Ni concentration which is equivalent to 218 μg g^{-1}. This is followed by Magotambis (*Syzygium longiflorum*) which belongs to the Myrtaceae family, and *Artocarpus sp.* which belongs to the Moraceae family at 142 μg g^{-1} and 123 μg g^{-1} of Nickel, respectively. None of these plants however qualifies as Ni hyperacumulators since hyperaccumulators should accumulate more than 1000 μg g^{-1} of dry weight. These plants however can still be considered as potential hyperaccumulators as compared to the other plants native to the area and are recommended plants suitable for the restoration of the mined-out areas of AMRI. These plant samples' exceptionally high nickel concentration compared to the other plant samples implies that they have the capability to tolerate high levels of Nickel and thus are recommended for the rehabilitation of the mined-out areas of AMRI.

Table 2. Ni content (μg g^{-1}) of screened plant species at AMRI Mined-out Sites

Plant Samples	Replication	μg g^{-1}	Mean μg g^{-1}	Description/Identity
S1P13	1	216	218	Common Name: Sowa-sowa
	2	222		Scientific Name: unidentified
	3	216		Family Name: Rubiaceae
S2P11	1	32	32	Common Name: Sowa-sowa
	2	32		Scientific Name: unidentified
	3	32		Family Name: Rubiaceae
S2P14	1	119	142	Common Name: Magotambis
	2	151		Scientific Name: *Syzygium longiflorum*
	3	157		Family Name: Myrtaceae
S4P3	1	49	60	Common Name: Balanti
	2	54		Scientific Name: *Homalanthus populneus*
	3	76		Family Name: Euphorbiaceae
S1P3	1	32	41	Common Name: Payuspos
	2	43		Scientific Name: *Leptuspermum flavenscens*
	3	49		Family Name: Myrtaceae
S4P5	1	119	123	Common Name: unknown
	2	119		Scientific Name: *Artocarpus sp.*
	3	130		Family Name: Moraceae
S4P16	1	32	29	Common Name: unknown
	2	27		Scientific Name: *Semecarpus sp.*

	3	27		Family Name: Anacardiaceae
S1P18	1	24	24	Common Name: unknown
	2	23		Scientific Name: *Syzygium sp.*
	3	26		Family Name: Myrtaceae
S3P7	1	27	27	Common Name: unknown
	2	27		Scientific Name: *Ficus sp.*
	3	27		Family Name: Moraceae
S4P22	1	23	23	Common Name: Molave/Tugas babae
	2	23		Scientific Name: *Vitex parviflora*
	3	23		Family Name: Verbenaceae

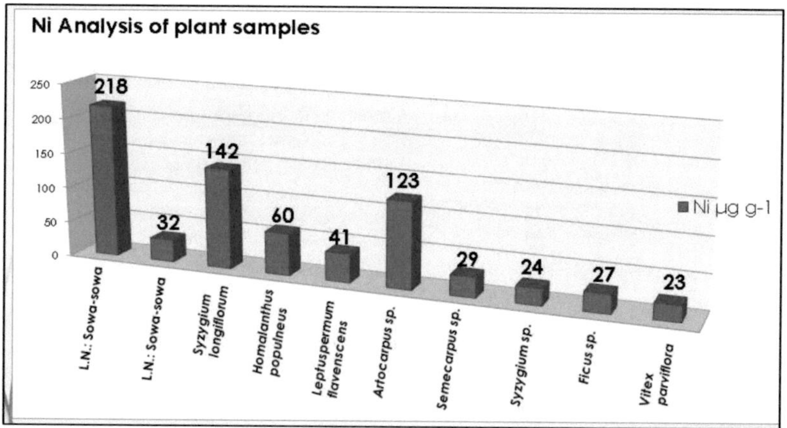

Figure 3. Comparison of total Nickel concentration ($\mu g\ g^{-1}$) of the plant samples after quantification

Implications

Phytoremediation presents a more environmental friendly, more ecological benefits and cost efficient alternative, with wide applicability compares to the traditional methods (Campos *et al.*, 2008; Mudgal *et al.*, 2010). Since the plant-based remediation is powered by solar energy, it is on average 20 times less expensive (Paz-Alberto & Sigua, 2013).

Also, the above plant species can be used in a technique called phytomining to grow and concentrate a metal. (Anderson *et al.*, 1999). Afterwards, the plant is harvested for metal extraction. Chaney (1983) as cited by Anderson et al., 1999 first suggested their use for phytomining. A phytomining operation would entail planting a hyperaccumulator crop over a low-grade ore body or mineralized soil, followed by harvesting and incineration of the biomass to produce a commercial bio-ore (Anderson *et al.*, 1999).

The use of indigenous or native plants would be effective to revegetate an area since these plants are well adapted to the local climate, insect and diseases (Wendy *et al.*, 2005 as cited by Edralin 2008), as well as the soil environment. These species if applied can serve as soil cover which prevents transport of heavy metals or toxic pollutants in the soil (Edralin, 2008). Furthermore, these plants would lead to the accumulation of soil organic matter, thus improving soil physical and chemical conditions and consequently improving the soil nutrient content which may in turn pave the way for the growth of other plant species and the establishment of soil organisms (Edralin, 2008).

Finally, Sowa-sowa (*Unidentified*), Magotambis (*Syzygium longiflorum*), and *Artocarpus sp.* are the most recommended plant species suitable for restoration (bioremediation principle) in the mined-out sites of Adnama Mining Resources Incorporated (AMRI) at Brgy. Urbiztondo, Claver, Surigao del Norte, Philippines.

LITERATURE CITED

ANDERSON, C.W.N., BROOKS, R.R., CHIARUCCI, A., LACOSTE, C.J., LEBLANC, M., ROBINSON, B.H., SIMCOCK, R., and STEWART, R.B. 1999. *Phytomining for nickel, thallium and gold.* Journal of Geochemical Exploration 67 (1999) 407–415. PII: S0375-6742 (99) 00055- 2.

BAKER, A.J.M.1995.*Metal hyperaccumulation by plants: our present knowledge of the ecophysiological phenomenon.* In: Will plants have a role in bioremediation? 14th Annual Symp. Current Topics in Plant Biochemistry, Physiology and Molecular Biology. Columbia, MO, pp. 7-8.

BAKER, A. J. M., and R., BROOKS. 1989. *Terrestrial higher plants which hyperaccumulate metalic elements.* A review of their distribution, ecology and phytochemistry. – Biorecovery 1; 81-126.

CAMPOS, V. M., I., MERINO, R., CASADO, L. F., PACIOS and L., GÓMEZ. 2008. Review. *Phytoremediation of Organic Pollutants.*Spanish Journal of Agricultural Research.Spanish Journal of Agricultural Research 2008, 6 (Special issue), 38-47 pp.

CULLAJ, A., HASKO, A., MCBOW, I., KONGOLI, F. 2004. *Investigation of the potential of several plants for phytoremediation of nickel contaminated soils and for nickel phytoextraction.*The European Journal of Mineral Processing and Environmental Protection Vol.4, No.2, e 1303-0868, pp. 144-15.

EDRALIN, D. I. A. 2008. *Copper, Lead, Nitrogen and Phosphorus Levels in Soils and Plants in the Abandoned Bagacay mine in Western Samar.* M.S.Thesis Manuscript, Visayas State University.

FERNANDO ES, QUIMADO MO, DORONILA AI. 2014. *Rinorea niccolifera (Violaceae), a new, nickel-hyperaccumulating species from Luzon Island, Philippines.* PhytoKeys 37: 1 – 13. doi: 10.3897/phytokeys.37.7136

LASAT, M. 2002.*The Use of Plants for the Removal of Toxic Metals from Contaminated Soil.* Innovative hazardous waste remediation. American Association for the Advancement of Science.

MUDGAL, V., N., MADAAN, and MUDGAL, A.2010.*Heavy Metals In Plants: Phytoremediation: Plants Used To Remediate Heavy Metal Pollution.*Agriculture and Biology Journal of North America. *<http://www.scihub.org/abjna>*.

PAZ-ALBERTO, A., AND G., SIGUA. 2013. *Phytoremediation: A Green Technology to Remove Environmental Pollutants.*American Journal of Climate Change, Vol. 2 No. 1, 2013, pp. 71-86. doi: *10.4236/ajcc.2013.21008.*

RASKIN, I., SMITH, R., SALT, D. 1997. *Phytoremediation of metals: using plants to remove pollutants from the environment.* Current Opinion in Biotechnology, 8:221–226.

SALAS, F.M., M. CHINO, S. GOTO, T. ISAGARASHI, H. MASUJIMA and K. KUMAZAWA. 1998. *Forms and Behavior of True Heavy Metals in Long Term Applied Soils with Sewage Sludges.* ISSASS Journal. Volume 4. No.1 pp: 60-97.

SALT, D.E., BLAYLOCK, M., KUMAR, N.P.B.A., DUSHENKOV, V., ENSLEY, B.D., CHET, I., 1995. *Phytoremediation: A novel strategy for the removal of toxic metals from the environment using plants.* Biotechnology, 13: 468-474

SCHMOGER, M.E., M. OVEN and E. GRILL. 2000. *Detoxification of arsenic by phytochelatins in plants.* Plant Physiol., 122, 793-80.

SHAW, A.J.1989.*Heavy metal tolerance in plants: Evolutionary aspects.* New York, USA, CRC Press 355 p.

Appendix Screening of plant species collected at AMRI Mined-out Sites for potential nickel hyperaccumulator. (Note: S indicates the sampling site while P indicates the Plant Sample shown under Plant Species column)

Plant Samples	Absorbance	Potential Rank	Plant Samples	Absorbance	Potential Rank
S1P13	0.6860	1st	S3P4	0.0320	42nd
S2P11	0.3094	2nd	S4P20	0.0320	43rd
S2P14	0.2871	3rd	S1P2	0.0283	44th
S4P3	0.2754	4th	S4P2	0.0280	45th
S1P3	0.1573	5th	S4P4	0.0260	46th
S4P5	0.1560	6th	S1P16	0.0226	47th
S4P16	0.1460	7th	S1P4	0.0214	48th
S1P18	0.1354	8th	S4P8	0.0190	49th
S3P7	0.1350	9th	S1P14	0.0182	50th
S4P22	0.1260	10th	S2P3	0.0154	51st
S3P5	0.1239	11th	S1P1	0.0151	52nd
S4P13	0.1200	12th	S2P9	0.0139	53rd
S4P19	0.1190	13th	S4P21	0.0120	54th
S3P6	0.1135	14th	S2P1	0.0107	55th
S4P11	0.0900	15th	S3P15	0.0104	56th
S2P5	0.0848	16th	S3P10	0.0102	57th
S4P25	0.0830	17th	S1P5	0.0095	58th
S2P13	0.0803	18th	S2P8	0.0091	59th
S4P26	0.0776	19th	S3P8	0.0090	60th
S4P24	0.0765	20th	S4P14	0.0090	61st
S4P18	0.0761	21st	S4P29	0.0085	62nd
S4P6	0.0760	22nd	S1P12	0.0084	63rd
S2P2	0.0734	23rd	S4P27	0.0080	64th
S3P2	0.0720	24th	S2P6	0.0078	65th
S3P14	0.0638	25th	S1P15	0.0076	66th
S1P7	0.0637	26th	S4P9	0.0068	67th
S4P10	0.0620	27th	S4P7	0.0065	68th
S4P23	0.0610	28th	S1P21	0.0063	69th
S3P3	0.0550	29th	S2P12	0.0060	70th
S3P12	0.0523	30th	S4P28	0.0058	71st
S4P12	0.0510	31st	S2P10	0.0054	72nd
S4P15	0.0510	32nd	S2P4	0.0052	73rd
S3P11	0.0480	33rd	S1P20	0.0036	74th
S1P19	0.0433	34th	S1P9	0.0031	75th
S1P8	0.0431	35th	S1P11	0.0027	76th
S4P17	0.0416	36th	S2P16	0.0019	77th
S4P1	0.0400	37th	S3P13	0.0014	78th
S1P17	0.0377	38th	S2P15	0.0008	79th
S3P1	0.0370	39th	S1P6	0.0003	80th
S1P10	0.0348	40th	S3P9	0.0002	81st
S2P7	0.0340	41st			